# MAKE it WORK!
# MACHINES

### Andrew Haslam

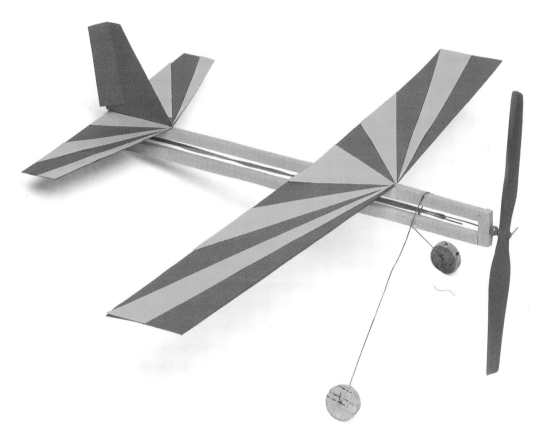

*written by*
David Glover  B.Sc., Ph.D.

*photography by*
Jon Barnes

**WORLD BOOK / TWO-CAN**

## MAKE it WORK!
## Other titles

Body
Building
Dinosaurs
Earth
Electricity
Flight
Insects
Photography
Plants
Ships
Sound
Space
Time

This edition published in the United States in 1997 by
World Book Inc., 525 W. Monroe, Chicago, IL 60661
in association with Two-Can Publishing Ltd.

**For information on other World Book products,
call 1-800-255-1750, x 2238.**

ISBN: 0-7166-4707-9 (pbk.)
ISBN: 0-7166-4706-0 (hbk.)

Printed in Hong Kong

1 2 3 4 5 6 7 8 9 10 99 98 97 96

Editor: Mike Hirst
Series concept and original design: Andrew Haslam and Wendy Baker
Additional design: Helen McDonagh

# Contents

Words that appear in **bold** in the text
are explained in the glossary.

# Being an Engineer

Humans are the only animals that invent and make machines. We use them to build skyscrapers, lift heavy loads, and move faster than the speed of sound. Humans have even made machines that can travel to the moon.

## MAKE it WORK!

You don't have to build robots or space rockets to be an engineer. In fact, engineers often make very simple machines. The projects in this book will show you how some machines are made, what they can do, and how they work. They'll also show you some engineering experiments. You might even be able to use the ideas for inventions of your own!

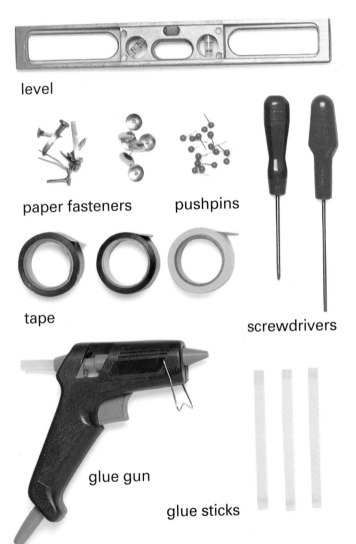

level

paper fasteners

pushpins

tape

screwdrivers

glue gun

glue sticks

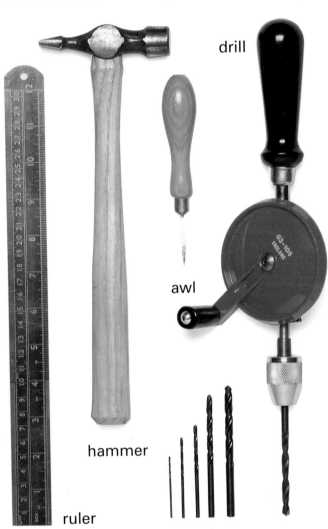

drill

awl

hammer

ruler

The scientists who build machines are called **engineers**. They do tests and experiments that help them to invent new machines and make old ones work better. Without engineers we wouldn't have tools or engines, trucks or trains, or even clocks or can openers.

## You will need

You can build most of your machines out of simple materials, such as cardboard and balsa wood, plastic bottles, and other odds and ends. However, you will need some tools to cut, shape, and join the different materials. All of the equipment shown above will come in very handy as part of an engineer's tool kit.

## Safety!

Sharp tools are dangerous! Always be careful when you use them, and ask an adult to help you. Make sure that anything you are cutting or drilling is held firmly so it cannot slip. A small table vise is ideal for holding pieces of wood.

## Planning and measuring

Always plan your machines carefully before starting to build. Measure the parts and mark them with a pencil before you cut. Mark the positions of holes before drilling them.

## Cutting

You will need saws for cutting wood and scissors for cutting cardboard and paper. A sharp craft knife is useful too, but be extra careful with the sharp blade.

## Drilling

To make some of the machines in this book, you will have to drill holes in pieces of wood. Use a pointed awl to start a hole and then finish it off neatly with a hand drill.

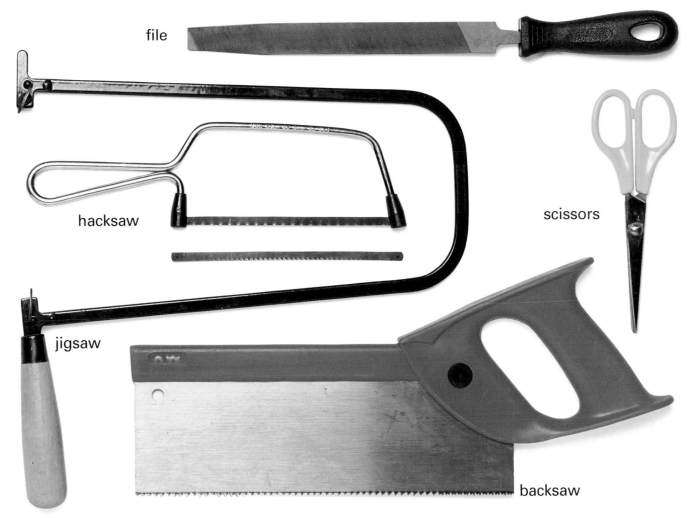

file

hacksaw

scissors

jigsaw

backsaw

## Joining

Strong glue is one of the simplest ways of joining pieces of wood, cardboard, or plastic. It's easiest to use glue sticks with a glue gun. Pushpins, paper fasteners, a hammer and nails, staples, and tape are all useful too. A level will come in handy if you want to make sure that parts are joined straight or level.

*Some machines are so simple that we don't always realize they are machines. But in fact, a machine is anything that applies a force to do a useful job. A pencil sharpener, for example, is a machine that uses a turning force to cut wood. Nutcrackers are machines that use a squeezing force to crack nuts.*

We often use machines to lift heavy weights or to help us move loads from one place to another. A wheelbarrow, for example, is a simple type of lifting machine. We use it to increase the **force** made by our muscles. If you had to move a pile of earth, you could carry a much heavier load in a wheelbarrow than you could lift in your own arms.

Perhaps the simplest machine of all for increasing force is the **lever**. A wheelbarrow is a kind of lever – and many other types of complicated machine are really just collections of levers that are put together to work in different ways.

## MAKE it WORK!

A simple lever is a straight rod that rests on a **pivot** or **fulcrum**. When you push one end of the rod down with an **effort**, the other end goes up, lifting the **load**.

Try making this model seesaw and find out for yourself how levers work.

**1** Mark the length of wood with paper strips spaced about 1 inch apart.

**2** Glue the dowel to the matchbox to make a pivot.

load

pivot or fulcrum

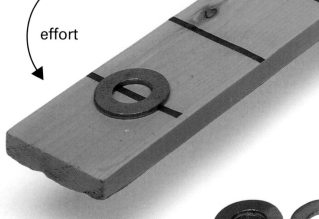

effort

**3** Place the center of the length of wood on the pivot so that the two ends balance.

Now try some experiments with the weights. Put a weight (the load) three marks from the fulcrum. Where must you place another weight (the effort) to lift the load?

## More load for less effort!

If the load is close to the fulcrum, it's easier to lift and you don't need so much effort. You may have noticed this if you've ever played on a seesaw – you can lift someone heavier than yourself if they sit nearer to the middle than you do.

Try putting two weights (the load) two marks away from the fulcrum of your seesaw. Where must you put a single weight to lift the load?

### To make a seesaw you will need

| | |
|---|---|
| a length of wood | glue |
| a small wooden dowel | a ruler |
| strips of colored paper | a pencil |
| weights (washers or coins) | a matchbox |

## Scales

Weighing scales use a balancing lever to make delicate measurements. Try building these scales. They are sensitive enough to weigh even a feather!

### You will need

cardboard and tape
thread and nails
a strip of wood
modeling clay
a short dowel
two glasses
pins

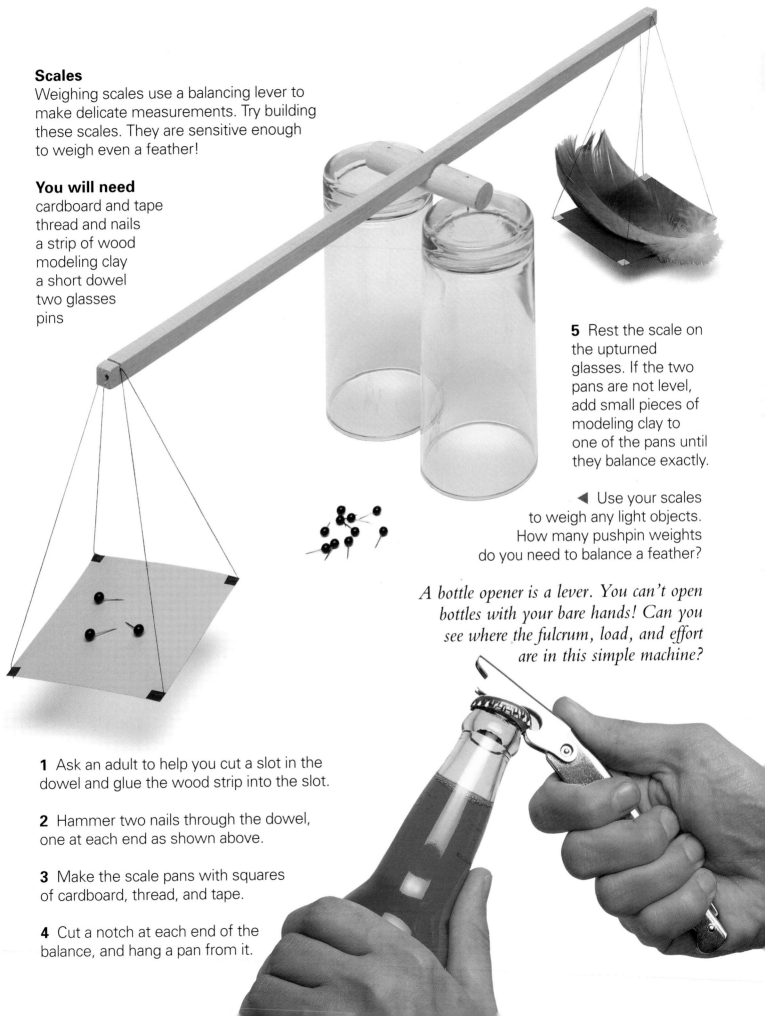

**5** Rest the scale on the upturned glasses. If the two pans are not level, add small pieces of modeling clay to one of the pans until they balance exactly.

◀ Use your scales to weigh any light objects. How many pushpin weights do you need to balance a feather?

*A bottle opener is a lever. You can't open bottles with your bare hands! Can you see where the fulcrum, load, and effort are in this simple machine?*

**1** Ask an adult to help you cut a slot in the dowel and glue the wood strip into the slot.

**2** Hammer two nails through the dowel, one at each end as shown above.

**3** Make the scale pans with squares of cardboard, thread, and tape.

**4** Cut a notch at each end of the balance, and hang a pan from it.

# Catapults

Have you ever flicked a pea from the end of a spoon? If you have, then you were using a lever. Your thumb was the pivot and your fingers applied the effort, making the bowl of the spoon move quickly and launching the pea up into the air.

A catapult works in just the same way. Before gunpowder was invented, ancient armies used catapults to fire rocks, burning rags, or other **projectiles** at their enemies.

**1** Ask an adult to help you cut the baseboard, the two side arms, and the main catapult arm from lengths of wood.

**2** Drill holes 1 inch apart along the main catapult arm and the side arms. The holes should be just big enough for the dowel to fit through. Before drilling, mark the position of each hole with an awl, as shown above.

**MAKE it WORK!**
The catapult on this page is powered by a stretched rubber band.

**3** Cut the triangular side pieces out of corrugated cardboard. You will need to use a sharp knife, so be careful not to cut yourself.

**You will need**
a small tin can or plastic cup
thick corrugated cardboard    paint
a thick rubber band    an awl
strong wood glue    pushpins
pieces of sponge    screw hooks
a hand drill    a wooden dowel
wood    a sharp craft knife

**4** Glue the cardboard side pieces and the wooden side arms to the baseboard. Then glue the small can or plastic cup to the end of the main catapult arm.

**5** Screw three hooks into the baseboard. Use the awl to mark the positions of the screws before you twist them in.

*When you fire an object from the catapult, it travels in a curved path called its **trajectory**. The distance the object travels is called its **range**. The range of the object and how high it goes depend on its speed and the angle from which it is launched.*

▶ This catapult experiment can be messy. Make sure you do it outdoors!

The catapult is designed so that you can change the position and height of the pivot (the wooden dowel) and the angle of the rubber band. How would you make the sponge travel the longest distance? How could you make it go as high as possible?

**6** When the glue is dry, you are ready to assemble the catapult. Line up a hole in the main arm with a pair of holes in the side arms. Push the dowel through the holes and hold it in place with a pushpin at either end.

**7** Tie the rubber band through the hole at the end of the catapult arm, and pull it over one of the hooks.

**Target practice**
Small pieces of sponge make good projectiles. If you soak the sponges in poster paint before you fire them, they will leave marks on a home-made cardboard target.

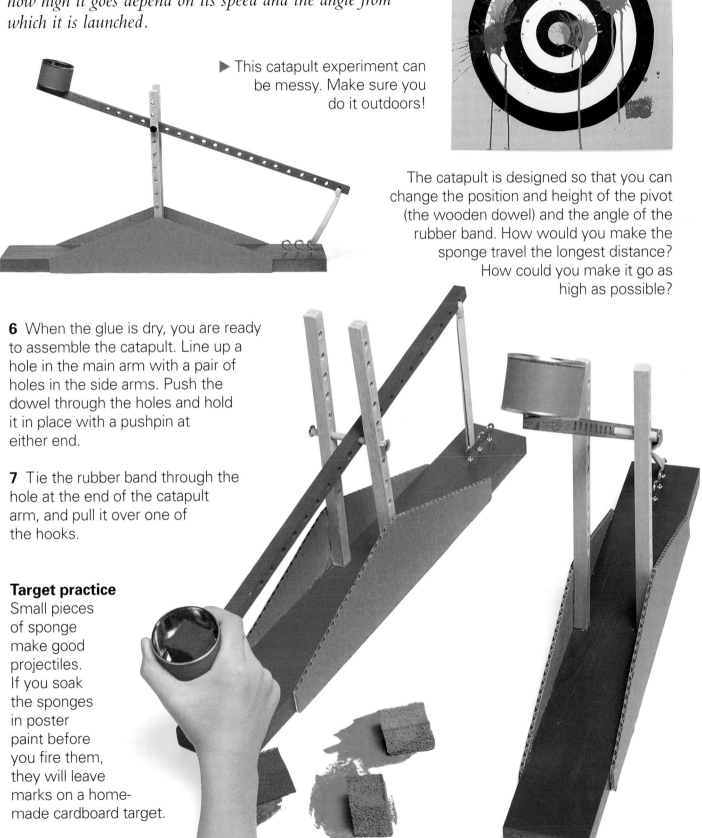

Lifting isn't the only job levers can do. We also use levers to change the direction of a movement.

The two ends of a simple lever always move in opposite directions. One end goes up when you push the other end down. By linking two levers with a flexible joint, we can make them move backward and forward as well as up and down. Bulldozers work in this way. The bones in our arms and legs are levers, connected at the knee and elbow joints.

**You will need**
glue
a cork
sandpaper
a hand drill
nuts and bolts
a pencil sharpener
lengths of balsa wood
a felt-tip pen and paper
a piece of wooden dowel
a drawing board and pushpins

## MAKE it WORK!

A pantograph is a drawing machine made from linked levers. Make one yourself and experiment to see how the linkages work.

**1** Cut two balsa wood lever arms 9 inches long and two more 5 inches long. Round the ends of the arms with sandpaper.

**2** Drill holes in the levers, just big enough for the bolts to fit through. Begin by making holes at the ends of all four levers.

**3** Now drill an extra hole in the middle of each of the two longer levers. If the pantograph is to work well, all the holes must be equally spaced – so be sure to measure carefully and mark the holes with a pencil before drilling.

**4** Join the longer levers at one end using a nut and bolt. Then attach a short lever to the middle of each long arm. Don't make the bolts too tight, since all the levers need to move freely.

**5** Ask an adult to help you drill a hole in the cork, just big enough for the pen to fit through.

**6** Glue the cork to the free end of one short lever. Push the pen through it and the free end of the other short arm as shown on the right.

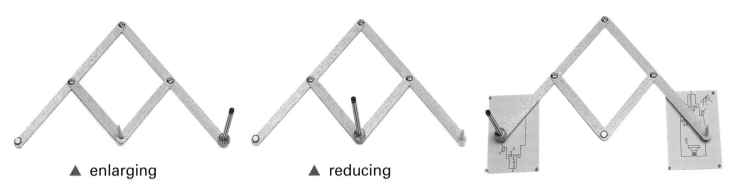

▲ enlarging      ▲ reducing

▲ drawing upside down

**7** Sharpen the piece of dowel to a point with a pencil sharpener and put it through the hole at the end of one of the long lever arms.

**8** Place the pantograph on a drawing board and push a pin through the last free hole.

**9** Pin the drawing you want to copy under the dowel pointer and put a blank piece of paper under the felt-tip pen.

**10** Trace around the drawing with the pointer and watch how the pantograph's lever arms carry the movement to the pen.

▼ With the pantograph set up like this, the copy is smaller than the original. The short arms are half the length of the long ones, so the copy is half the size.

### Different drawings

If you swap around the positions of the pin, wooden pointer, and felt-tip pen, you can also make the pantograph draw larger, or the same size but upside down. The photographs above show how you should arrange the parts to produce some of these different results with your pantograph.

### Pantograph experiments

What do you think would happen if you changed the length of the lever arms? You could experiment by drilling more holes along the arms and then bolting them together in different ways. You may end up with drawings that look stretched, squashed, or tilted!

*The movements made by linked levers depend on two things: the length of the levers and the positions of the joints. A movement can be made bigger or smaller simply by changing the way the lever rods are linked to one another.*

Imagine that you wanted to hang a flag from the top of a tall pole without moving your feet from the ground. How could you do it?

The easiest answer would be to use a **pulley**, attached to the top of the pole, with a rope looped over it.

A pulley changes a downward pull on one end of a rope into an upward pull at the other end. With simple pulleys we can lift all kinds of loads up poles or tall buildings – and if you have a window blind at home, you'll be using pulleys yourself every time you pull it up or down.

## MAKE it WORK!

Thread spools make first-class pulley wheels. With a few spools you can make a whole set of pulleys to experiment with.

### You will need

empty thread spools
string or cord
yogurt cups
thick wire
eyebolts
pliers
sand

### ▲ Single pulley

**1** Push a piece of wire through the hole in a thread spool. Use the pliers to cut, bend, and twist the wire to make a pulley as shown.

**2** Make sure the thread spool spins freely on the wire and then hang the pulley from an eyebolt.

**3** Loop the string over the pulley and tie a load to one end. A yogurt cup filled with sand is a good load. Push a wire through the sides of the cup to make the handle.

## Easing the load

Experiment to see how easily you can lift the cup full of sand with your simple pulley. One single pulley won't make it any easier to lift the load, it just changes the direction in which you apply the force. You pull down on the rope to make the load go up. With a single pulley you cannot lift anything heavier than you could using just the strength of your arms. However, see what happens if you use two or more pulley wheels together.

### ◀ Double pulley

**1** Make a second simple pulley just like the first, and hook it to the cup handle.

**2** Tie one end of the string to the top of the wire hanger on the first pulley as shown.

**3** Loop the string under the lower pulley and then back up over the top of the upper pulley.

Now test the double pulley system to feel how difficult it is to lift a load. Do you need to use more force than with a single pulley, or less?

### ▶ Quadruple pulley

To make a pulley system with four wheels you will need to make two twin **pulley blocks**.

**1** Using a longer piece of wire, make two new hangers, each wide enough to hold two thread spools, positioned side by side.

**2** Tie the string to the top hanger. Loop it down under one of the lower pulleys and then around each of the other pulleys as shown.

How does this system of four pulley reels compare to the simpler ones?

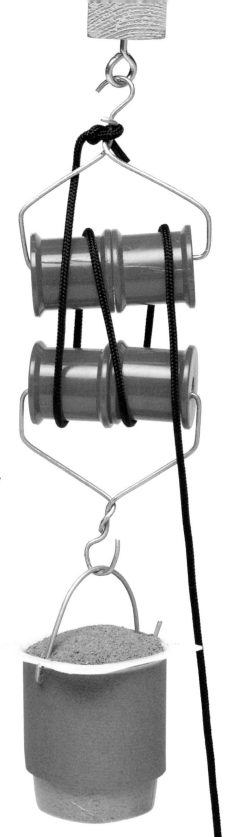

## Heavier weights, shorter distances

With two pulleys you can lift almost twice as much as with a single pulley, without using any extra force. But you don't get anything for nothing – the load only travels up half as far as the distance you pull on the string! Four pulleys lift almost four times as much weight.

*Pulley systems work in about the same way as levers. They help us to lift big loads with just a small effort. With a pulley block, a car mechanic can lift the engine out of a car in order to repair or replace it.*

Have you ever been ice-skating? Skates glide smoothly across the ice and you move with hardly any effort. Rubber boots, on the other hand, are not slippery at all. They keep a firm grip on the ground, and keep you from sliding even if you're walking in slippery mud.

▲ Put some marbles under a pan lid on a smooth surface. The marbles cut down the friction, and the lid rolls around smoothly. Balls that reduce friction in this way are called **ball bearings**.

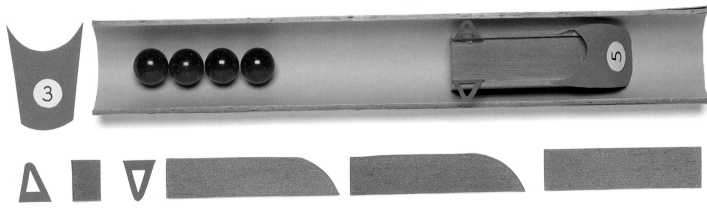

## Friction

Rubber boots grip the ground firmly because of **friction**. This is an invisible force, caused when two objects rub against one another. Friction keeps things from sliding. When rubber rubs against anything, it causes lots of friction. But thin strips of metal on ice make hardly any friction at all.

## Bearings

Friction can be a nuisance in machines, and it may keep the parts from moving smoothly. Many machines contain ball bearings to cut down on friction. There are ball bearings inside the **hub** of a bicycle wheel. As the wheel rotates, the steel balls turn around inside the hub.

## MAKE it WORK!

At the Winter Olympics, bobsleds hurtle down the icy bobsled run at thrilling speeds. These model bobsleds don't run on ice – but they can still pick up plenty of speed as they race down their tracks of cardboard.

### You will need

strong wood glue
cardboard tubes
wooden dowels
thin cardboard
some marbles
modeling clay
balsa wood

**4** Build a track from sections of tube that are connected with curves of fairly thin cardboard. Hold the track up on pieces of wooden dowel connected with lumps of modeling clay.

**5** Decorate your track with cardboard flags and colored markers.

**6** Place the bobsled on the marbles and set it off down the run. The marbles don't make much friction, so the sled picks up speed and will be going fast once it reaches the bottom!

### Bobsled races

Build a double run, make two bobsleds, and you can hold competitions. Try adding weights to the sleds (use lumps of clay) – do the weights make the sleds go faster?

*Ball bearings aren't the only way of cutting down friction. Oil is a good solution, too. The slippery liquid spreads out in a very thin layer between the moving parts of a machine. Oil is a vital part of most* **engines***.*

**1** Cut out the balsa wood pieces to make the sides, top, and back of the bobsled. The bobsled should be slightly wider than the marbles, but not quite as deep.

**2** Glue the balsa wood parts together. Cut the nose and the tail fins from thin cardboard and glue them in place.

**3** Ask an adult to help you cut some cardboard tubes in half lengthwise. (The insides of old aluminum foil rolls come in handy here, or, if you have some old plastic gutters, you could even use them to make the bobsled track.)

Merry-go-rounds, sewing machines, record players, fishing reels, washing machines, and bicycles: these are just a few of the many machines that turn, or **rotate,** as they work.

All the different rotating parts inside a machine can be connected with a **drive belt.** As one part turns, it drags the belt around with it, carrying its turning motion to all the other parts of the machine.

**1** Cut the sandpaper into strips, and glue a strip around each of the thread spools. The rough surface of the sandpaper is needed to make some friction between the reels and the belt. This way, the belt will not slip.

**2** Draw both the front and the back of each acrobat on a piece of cardboard as shown, leaving a space between front and back to make a base. Cut out the figures. Then fold and glue them so that they stand up.

**3** Glue an acrobat onto each spool.

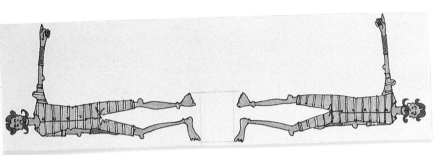

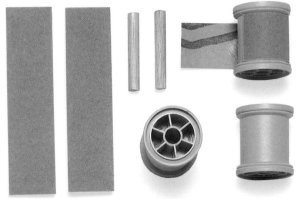

### MAKE it WORK!
A drive belt runs around a series of pulleys to carry the turning force from one place to another. If the belt is going to work properly, there must be **friction** between it and the pulleys, so that the belt does not slip. If the belt is too slack, it will not grip. If it is too tight, it might break or twist the pulleys out of line. These whirling acrobats stand on thread spools connected by a belt made of ribbon.

### You will need
sandpaper
a wooden board
glue and cardboard
a piece of hook-and-loop fasteners

empty thread spools
a wooden dowel
a ribbon

**4** Ask an adult to help you cut the wooden dowel into a number of shorter dowel pegs. Smooth the ends of the pegs with sandpaper.

**5** Drill holes into the baseboard. They should be just big enough for the dowel pegs to fit snugly into them.

**6** Put the pegs into the holes, and then put a thread spool onto each peg. Make sure that every reel can turn freely on its peg.

**7** Push a short piece of dowel into the gap between the center hole and the rim of one thread spool. This is the drive belt handle. You will use it to turn the drive belt.

**8** Stretch a length of ribbon around the spools so that it touches them all. Use a piece of loop-and-hook fastener material to join the ends of the ribbon. Then you can adjust it so that it is not too tight and not too slack, and the acrobats will turn more smoothly.

## Changing direction
Both of the acrobats on the top board move in the same direction when you turn the handle. But on the bottom board, two acrobats turn in a direction opposite of the other four. Can you thread your drive belt so that some acrobats turn one way and some the other?

*Old cars had to be started by turning the engine with a crank handle. When the engine was going, it was kept cool by a fan turned by a rubber belt. If the fan belt snapped, the water in the engine boiled over. Modern cars usually have electric starters and electric fans.*

## Changing speed
If all the thread spools have the same diameter, they all rotate at the same speed. But if you use different-sized spools, they turn at different speeds. To turn a big spool, the belt has to move farther than it does to turn a smaller one, and so the big spool turns around more slowly.

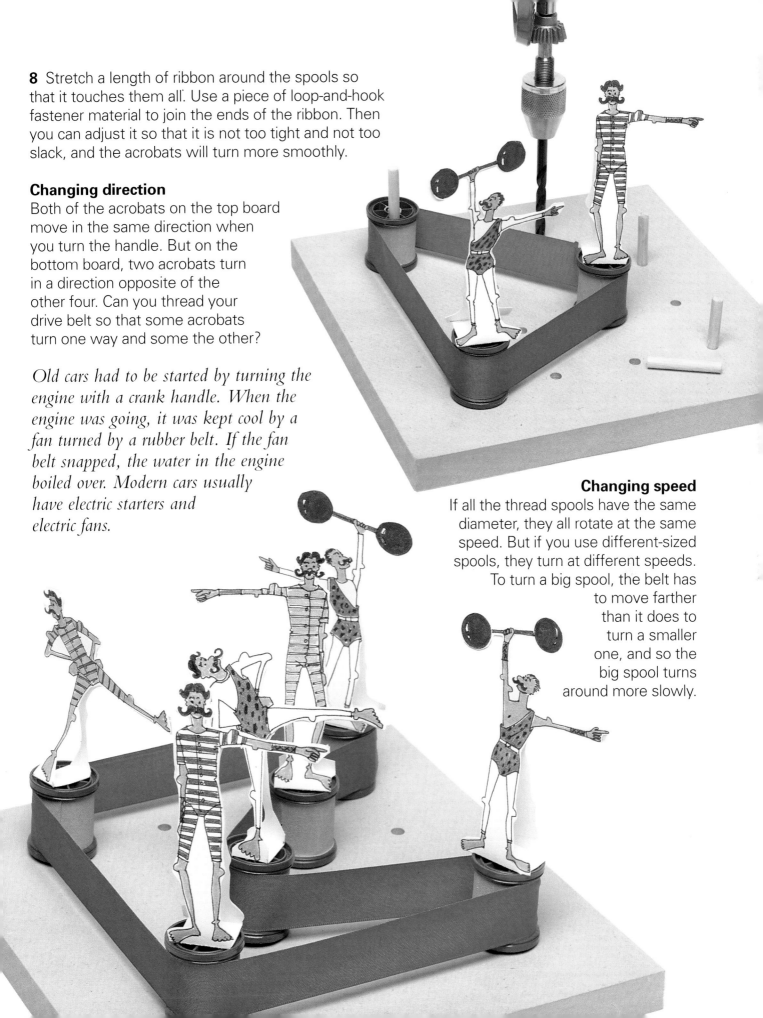

You'll find **gears** inside nearly every machine that turns. Clocks, watches, and bicycles all use them. Just like a belt drive, the gears connect all of the rotating parts, but gears last longer than belts and are more precise. If you've ever ridden a mountain bike, you'll know that gears are a good way of changing speed too.

## MAKE it WORK!

The best way to find out how gears work is to make some of your own to experiment with. Each of these homemade gears is made from a jar lid with a strip of corrugated cardboard stuck around the rim. The corrugations face out to make the gear teeth.

**1** Bend a strip of cardboard around the rim of a jar lid. Try to stretch it into place so that there is a whole number of teeth evenly spaced around the lid. Cut the strip carefully to length and then glue it in place.

### You will need

strips of corrugated cardboard about ½ inch wide (You can make these by peeling apart the thick cardboard sides of a cardboard box.)
jar lids and bottle tops of different sizes
a pin board and pushpins
a short dowel peg
glue and paper

**2** Make a small hole in the middle of the gear and pin it to the board so that it spins freely.

**3** Make a selection of different-sized gears to add to the board. Glue a dowel peg to one of the gears to make a crank handle.

**4** To make the gears work you must place them so the teeth **mesh**. When you turn one gear its teeth will push on its neighbor's teeth and make them turn in the opposite direction.

## Gear experiments

Connect a series of gears like the one shown above. If you turn the big gear, what happens to the two smaller ones? Which way do they go around? Which does a complete turn first?

Now try turning the small gear – do the bigger gears turn more quickly or more slowly?

Count the number of teeth on each gear. If you turned a gear with 20 teeth around once, how many times would it turn a gear with 10 teeth?

## Drive chains

In some machines, gears called **sprockets** are connected by a **drive chain**. A bicycle chain connects a sprocket on the pedals to another one on the back wheel. The chain transfers the movement from the pedals to the wheels.

▼ Make a model drive chain from a long strip of corrugated cardboard with the ends taped together. Loop it around two different-sized gears and work out how far the small gear moves when you turn the larger one.

Turning gears make beautiful patterns. As a gear goes around, each point on the wheel's surface follows a different path. By tracing these paths, we can draw patterns of loops and curves that repeat and shift as the gear rotates.

**1** Use a craft knife to cut a large, circular hole in a square of stiff cardboard.

**2** Glue a narrow strip of corrugated cardboard around the inside of the hole. Position the strip so that the corrugations face into the hole to make gear teeth. Make sure that one edge of the strip is level with the edge of the hole, so that the cardboard will lie flat on top of the pin board.

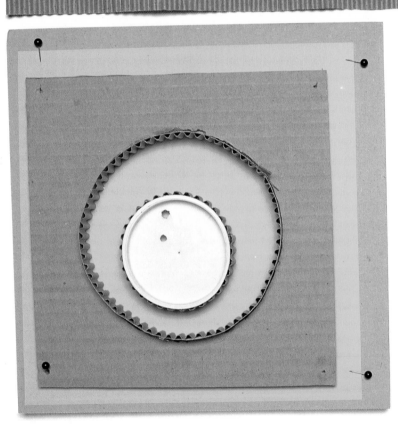

### MAKE it WORK!
This gear drawing machine uses the homemade lid and corrugated cardboard gears from the page before. Experiment with different-sized gears to discover all kinds of patterns.

### You will need
pushpins
a pin board
corrugated cardboard
gears made from lids and cardboard (Plastic lids are easiest to make holes in.)
drawing paper
a pen or pencil

**3** Put a sheet of paper on the pin board and then pin the square with the hole in it on top.

**4** Make small holes in your gear wheels at different distances from the center. The holes must be big enough for the point of your pen to fit through.

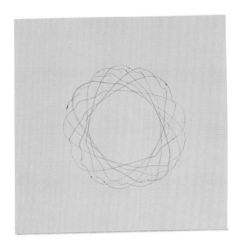
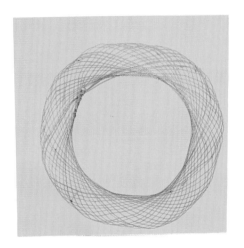

**5** Place a gear wheel in the large cardboard hole and put the point of a pen through one of the small holes so it touches the paper.

**6** Hold the board firmly with one hand and use the pen to push the gear carefully around the inside of the large circle. As the gear rotates, the pen draws a line on the paper.

**7** Keep pushing the pen and the gear around to build up a beautiful curved pattern.

▲ Try to make some drawings using different-sized gears and with holes that are at different distances from the center of the gear.

Some patterns will repeat after just a few turns, others may take many turns before they start again. Think about how the number of teeth on the gears and the position of the pen hole affect the pattern. You could also make up a second gear board with a different-sized hole to investigate even more patterns.

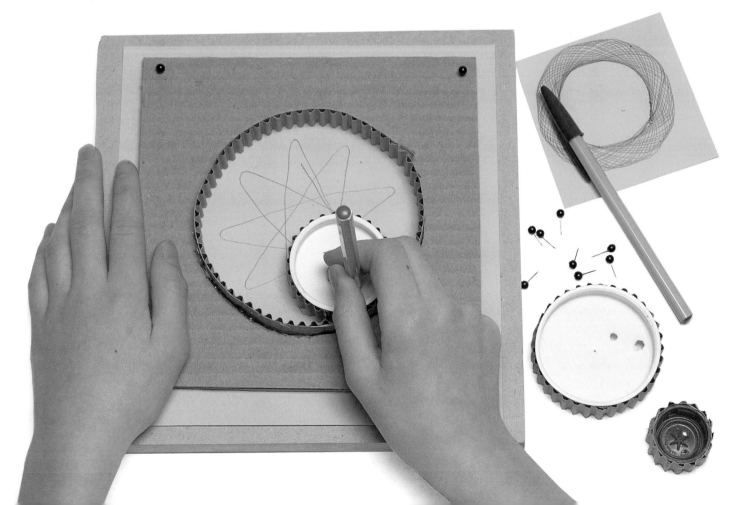

Gears let us transfer movement from one wheel to another, but how do you change a turning movement into an up and down one? The answer is a **cam**.

A cam is like a wheel, but with the **axle** (the shaft that goes through the middle) off center. If you watch a point on the edge of the cam, it seems to move up and down as the axle of the cam goes around and around.

## MAKE it WORK!

This model camshaft shows exactly how rotating cams can move things up and down – and in the order you choose.

### You will need

wood  
strong wood glue  
two thick cardboard tubes of different widths  
a small tack  
thick and thin dowels

**1** Ask an adult to help you cut the pieces of wood needed for the frame which will hold the camshaft and the four plungers. The height of the frame must be at least twice the diameter of the cams. The plunger tubes must fit tightly into the space across the top of the frame.

**2** Cut the wide cardboard tube into four rings. These will be your cams.

**3** Cut eight short strips of wood that just fit across the cardboard cams. At the end of each strip, drill a hole with a diameter a little larger than that of the thin dowel rod.

**4** Cut slots in the rims of the cams and glue the wooden strips in place as shown. Make sure the holes face each other on opposite sides of each cam.

**5** When the glue is dry, push the cams onto the dowel rod to complete the camshaft.

**6** Cut four pieces of the narrow cardboard tube and four longer pieces of the fatter dowel to fit inside. These will be the plungers.

**7** Drill two holes, facing one another, halfway down each of the frame's side pieces.

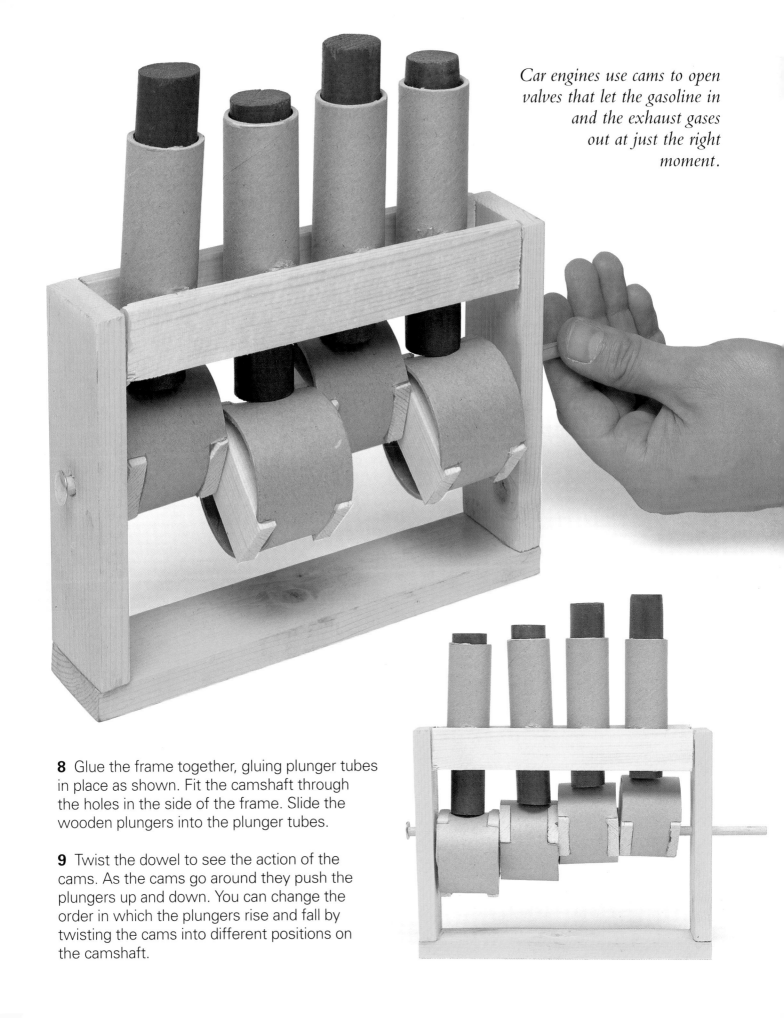

*Car engines use cams to open valves that let the gasoline in and the exhaust gases out at just the right moment.*

**8** Glue the frame together, gluing plunger tubes in place as shown. Fit the camshaft through the holes in the side of the frame. Slide the wooden plungers into the plunger tubes.

**9** Twist the dowel to see the action of the cams. As the cams go around they push the plungers up and down. You can change the order in which the plungers rise and fall by twisting the cams into different positions on the camshaft.

Everything on Earth is affected by an invisible force called **gravity**. The force of gravity causes objects to fall toward the ground. If you drop a tennis ball from the top of a tall building, gravity makes it fall to the earth below.

If you jump from a diving board, gravity causes you to fall straight down into the swimming pool, but if you set off down a water slide, you will go down only at the same angle as the slide. The steeper the slope, the faster you go!

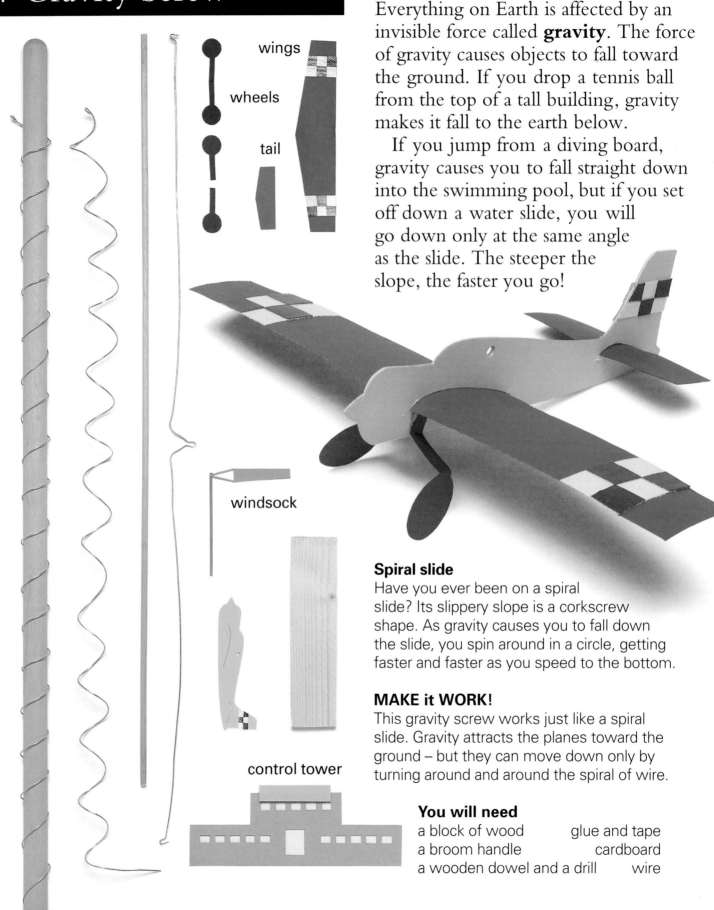

wings

wheels

tail

windsock

control tower

## Spiral slide

Have you ever been on a spiral slide? Its slippery slope is a corkscrew shape. As gravity causes you to fall down the slide, you spin around in a circle, getting faster and faster as you speed to the bottom.

## MAKE it WORK!

This gravity screw works just like a spiral slide. Gravity attracts the planes toward the ground – but they can move down only by turning around and around the spiral of wire.

### You will need

| | |
|---|---|
| a block of wood | glue and tape |
| a broom handle | cardboard |
| a wooden dowel and a drill | wire |

**1** Wind a length of wire around the broom handle to make the spiral. Take care to make the loops of the spiral evenly spaced. Slide the finished spiral off the handle.

**8** Hang the planes on the hooks and put the hanger on the top of the spiral. Let it go, and see if it runs smoothly down the spiral. You may need to adjust the shape of the hanger loop and some of the twists in the spiral to get the planes to fly really well.

**2** Ask an adult to help you drill a hole in the block of wood. Glue the dowel in the hole.

**3** Slip the wire spiral over the dowel. Use a piece of tape to hold it in place on the block.

**4** Make a cardboard windsock and control tower, and then glue them to the block.

**5** Cut out the parts for the two cardboard planes. Make two slits in the body and slide the wings and tail through each. Glue the wheels under the wings.

**6** Gently hold the body of one plane between your thumb and finger to find the place where it balances. Make a small hole in the body at this point.

**7** Bend a second length of wire into a hanger shape. Make a loop in the middle to fit over the wire spiral and bend small hooks at each end.

*The seeds of the sycamore tree grow in pairs. Each seed has a wing attached, which makes every pair spin around as it falls out of the tree. The spinning movement slows down the fall of the seeds, so that they catch the wind and travel farther away from the tree before finally settling on the ground.*

Spirals and screws come in handy for moving things up as well as down. One of the first people to use a screw as a lifting machine was the ancient Greek scientist Archimedes. He invented a screw pump that could raise water from a lower level up to a higher one, making it flow against the force of gravity.

**3** Cut a small hole in the center of each disk, the same diameter as the wooden dowel.

**4** Make a slit in each cardboard disk, from the center hole to the edge.

**5** Now join the cardboard disks to make the screw shape. Take two disks. Glue the edge of the slit in one disk onto the opposite edge of the slit in the second disk.

**6** Next, glue the free edge of the slit in the second disk to the opposite edge of the slit in a third disk.

**7** Continue to glue the slit edges in this way until all six disks have been stuck together to make a screw.

**8** Push the dowel through the holes in the centers of the disks, and stretch out the screw along the length of the dowel as shown below. Glue the two free ends of the cardboard screw firmly to the dowel.

## MAKE it WORK!
This model Archimedean screw is not really strong enough, or made of the right materials, to lift water, but it is an ideal dispenser for popcorn or breakfast cereal.

### You will need
a wooden dowel
a sharp craft knife
a plastic soft drink bottle

glue
a small tack
stiff cardboard

**1** Ask an adult to help you cut the bottom off the bottle and to cut a triangular hole in the neck as shown.

**2** Cut out six cardboard disks, just big enough to fit inside the plastic bottle.

**9** Slide the completed screw into the bottle. Hold it in place with a small tack pushed through the bottle cap and into the end of the wooden dowel.

**10** Now test your Archimedean screw. Dip the bottle into a bowl of popcorn and twist the dowel gently with your fingers to draw some popcorn up out of the bowl.

*Archimedean screws are used in combine harvesters to lift grain into storage containers.*

▲ Just like cams, screws are a way of changing one kind of movement into another. Our popcorn dispenser changes rotation (twisting the dowel) into upward movement.

*Although Archimedean screws were first built more than two thousand years ago, they are still used today. In some parts of Africa, farmers use them for irrigating their crops. The screws lift water out of rivers into raised irrigation canals. These ancient water pumps are powered either by animals or by hand.*

# 28 Balance

Because gravity attracts things toward the ground, it is much simpler to stay on a trapeze than it is to balance on a high wire. A trapeze artist's weight hangs below her hands, so as long as she has strong arms and holds on tight, she won't fall off. But a tightrope walker's weight is all above her feet – she only has to lean over a little and she topples from the wire.

## MAKE it WORK!

Try making these simple balancing toys. They seem to stand above the wire, but they work because really most of their weight is hanging below it.

### To make balancing acrobats you will need

| | |
|---|---|
| stiff cardboard | colored pens |
| small metal washers | glue and scissors |

**1** Draw an acrobat on a piece of cardboard. The left-hand side should be a mirror image of the right.

**2** Cut out the acrobat. Cut a small notch in his hat where he will balance on the wire.

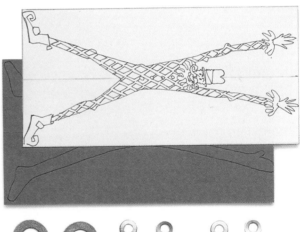

**3** Glue a washer to each of the acrobat's hands.

The weight of the washers below the wire helps the acrobats to balance above it.

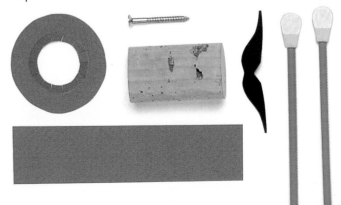

▲ A **pendulum** bob hangs with its weight as low as possible. If you push it aside, the force of gravity brings it back again.

*Some tightrope walkers carry a long, flexible pole. It helps to keep their weight low, just like the mustache men's knitting needles.*

## Gyroscopes

A gyroscope is a machine that seems to defy the force of gravity. It has a heavy metal disk, which spins around on an axle inside a frame. Although gravity still attracts the machine down, the spinning movement of the disk stops it from toppling over, and so the gyroscope balances on the wire.

◀ balancing acrobats

▶ gyroscope

◀ mustache men

### To make mustache men you will need

knitting needles
brass screws
cardboard
scissors
corks
glue

**1** Cut out the cardboard mustache and hat pieces. Glue them on the cork as shown.

**2** Twist a screw into the bottom of the cork.

**3** Ask an adult to help you push two knitting needles into the cork at an angle.

Just like the acrobats, the mustache men balance because most of their weight is in the knitting needles, below the wire.

▶ Setting a simple gyroscope spinning.

*A gyroscope has a very useful feature – once it is spinning, the axle will keep on pointing in the same direction as long as it is allowed to move freely. In the early twentieth century, scientists used this feature to develop a new kind of compass – the **gyrocompass**, which is used in most ships and aircraft today.*

**Pneumatic** machines use air to transfer force from one place to another. We tend to think that air is weak and thin, but if it is squashed together, or **compressed**, it can push with tremendous strength. A hurricane, for instance, can blow down trees and buildings. The air inside an air mattress will hold up the weight of a person. And a tire filled with air can carry the weight of a huge truck or a jumbo jet.

## MAKE it WORK!

This pneumatic man is fired by squeezing the plastic bottle to compress the air inside. The air is pushed out along the straw and launches the flyer like a human cannonball.

**You will need**
a liquid detergent bottle
cardboard and thread
a plastic bag
a thin straw
a fat straw
tape

**1** Seal one end of the fat straw with a piece of tape.

**2** Cut out the shape of the man in thin cardboard and stick him to the sealed end of the straw.

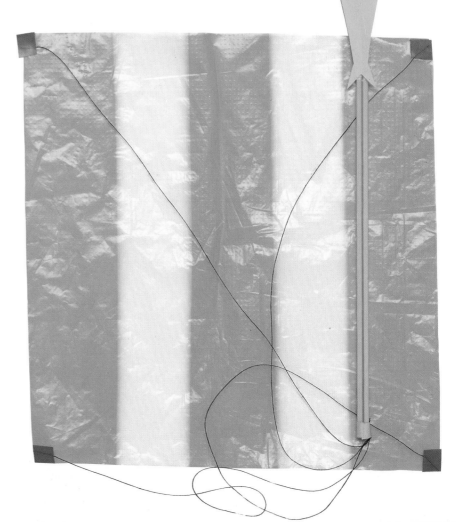

**3** Push the thin straw onto the nozzle of the dishwashing liquid bottle. If it does not fit very well, seal it in with modeling clay or glue.

**4** Slide the fat straw over the thin straw. If the fat straw doesn't fit neatly, cut a slit up its side and pull it a little tighter around the thinner straw. Then seal it up again with tape.

**5** To test the pneumatic man, squeeze the bottle sharply. The compressed air inside the bottle pushes against the sealed end of the thick straw as it tries to get out. The flying man is launched along a curved **trajectory**, like the sponges launched from the catapult on page 9.

### Parachute
If your man is a highflier, you could equip him with a parachute so that he has a soft landing.

**1** Cut a 7 inch square sheet from a plastic bag.

**2** Tape an equal length of thread to each corner of the plastic square. Then tape all the free ends of thread to the base of the fat straw.

**3** Fold the parachute into a strip and lay it alongside the straw.

**4** Launch the man in the usual way. The parachute will unfold and bring him gently back down to the ground.

▶ As it comes down, the open parachute fills with air. Air pushing upward underneath the plastic slows down the man's fall, just as it would slow the fall of a tissue or feather.

### Pneumatic tires
Pneumatic tires are tires filled with compressed air. Before these tires were invented, carts and bicycles had simple tires made from solid rubber strips. Pneumatic tires are a great improvement because they are springier, so they give a much more comfortable ride than solid rubber.

*Pneumatic tires were invented by John Dunlop in 1888. He had the idea for them when he saw his son riding a tricycle over a piece of rough ground. Dunlop made his first air-filled tire from a length of rubber garden hose. The company he founded still makes tires today.*

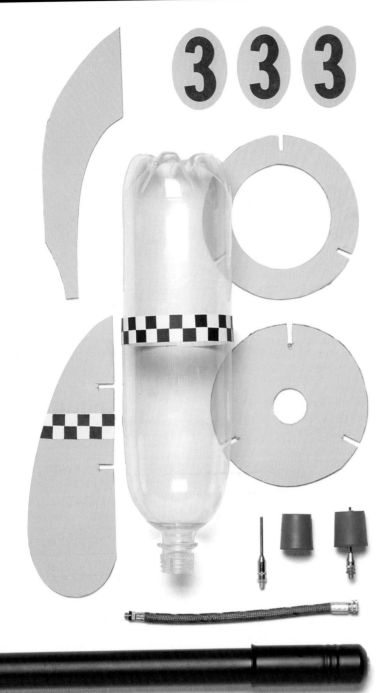

All machines need **energy** to make them go. Our pulleys were turned by human muscle energy. The gravity screw worked by the downward force of gravity on the plane. But most big machines today are driven by **engines**.

An engine makes power by burning a fuel such as gasoline or coal. Burning the fuel releases the energy it contains. A rocket engine works by burning the rocket fuel so that it squirts hot gases backward at great speed. As the gases push back, the rocket is thrust forward and shoots up into the sky.

### MAKE it WORK!

This water rocket isn't powered by rocket fuel, but it does work in a way similar to a real rocket by using just air and water. The space above the water is pumped full of compressed air with a bicycle pump. Eventually, the energy stored in the squashed air pushes the water out of the base, and the rocket is thrust up off the ground.

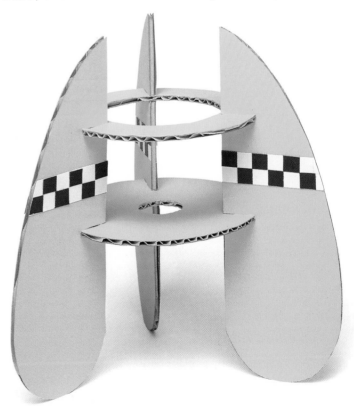

### To make a water rocket you will need

strong glue
a bicycle pump
a plastic bottle
an air valve (The kind that is used for blowing up footballs is best – you can buy one at a sporting goods store.)

tape
a rubber stopper or cork
thick, corrugated cardboard

**Be very careful!**
This rocket is very powerful and could hurt people seriously if it hit them. **Never** launch it without an adult to help you.

- **Always** fly the rocket out of doors in a wide empty space, well away from roads.
- **Never** fly the rocket near other people.
- **Don't** stand over the rocket as you pump it up. Keep off to the side.

**1** Cut the three base fins, two base rings, and three nose cone parts from corrugated cardboard.

**2** Make the rocket base from the fins and the two rings as shown. Stick the parts together with tape or strong glue. Then glue the base onto the plastic bottle.

**3** Make the nose cone and attach it to the top of the rocket.

**4** Ask an adult to help you make a small hole through the rubber stopper with a pin or a skewer. Then push the air valve through the stopper.

**5** Choose your launch site carefully. (See the safety note above.)

**6** Pour water into the bottle until it is about one-third full. Push the stopper tightly into the neck of the bottle and stand the rocket on its base. Attach the bicycle pump to the air valve, stand off to the side, and start pumping.

◀ As you pump, you will see the bubbles of air rising through the water. The pressure builds up inside the bottle until the stopper can no longer hold in place. Suddenly, the rocket blasts off, squirting out water as it lifts into the sky.

A windmill uses the force of the wind to do useful work. Waterwheels turn the energy of running water into useful power. Before the first steam engines were invented, windmills and water-wheels were almost the only machines that were not powered by human or animal muscles. Farmers often used them to grind corn and pump water.

## MAKE it WORK!
Try making this simple windmill. The wind turns a crank, which makes a rod go up and down.

### You will need
| | |
|---|---|
| a wooden dowel | wood |
| strong wood glue | a drill |
| a sharp craft knife | a cork |
| strips of thin plastic | cardboard |
| a tube of thick cardboard | wire and beads |

**1** Cut four strips of wood to length to make the frame. Bend the wire crank to shape.

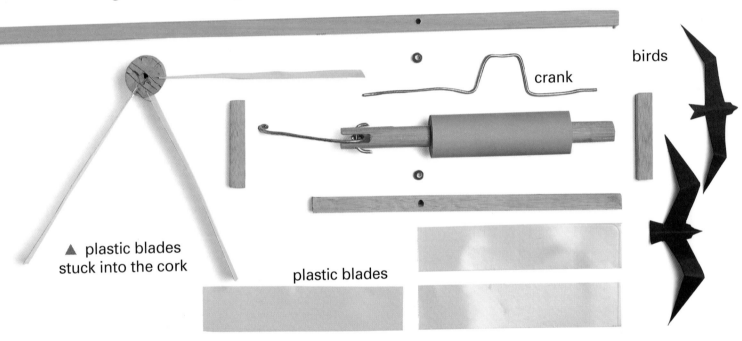

crank

birds

▲ plastic blades stuck into the cork

plastic blades

When engines fueled by coal and oil came along, windmills and water-wheels began to disappear. However, they are now becoming popular again. Today, we are more aware of the **pollution** caused by burning **fossil fuels** such as coal, gas, and oil. In comparison, wind and water power are clean and quiet sources of energy. They have another advantage, too – unlike coal, gas, and oil, our supplies of wind and water will never run out!

**2** Drill holes facing each other in the longer frame side pieces to take the crank. Glue the frame together with the crank in place.

**3** Ask an adult to help you cut slits in the cork. Slip the blades in place and secure with glue.

**4** Slip a bead over each end of the crank shaft. Then push the cork onto one end of the shaft. Bend the other end to keep the shaft in place.

**5** Cut a slot in one end of the dowel rod. Drill a hole at right angles to the slot and pass a small horseshoe-shaped wire through the hole.

**6** Glue the cardboard tube to the top of the frame and push the dowel through it. Connect the wire horseshoe to the crank with a third piece of wire as shown. Make sure the dowel moves up and down easily as the crank turns.

**7** Glue cardboard birds to the top of the dowel.

▶ Turn the mill by blowing a hair dryer at the blades. Experiment with the windmill out of doors too. Does it catch the wind better if you fit the blades into the cork at a different angle?

*Modern windmills don't just grind corn or pump water. Nowadays, engineers can also build windmills to generate electricity.*

## Water power
Waterwheels can be built wherever there is fast-flowing water that will turn the blades of the wheel. Most modern waterwheels are complicated machines that are used to make electricity. They are called **hydroelectric turbines** and the electricity they produce is **hydroelectricity**. Hydroelectric turbines are usually built along big rivers or in dams, where water is made to pass through a turbine in order to get out of a reservoir. Electricity can even be generated in coastal areas by the movement of the tides through a turbine.

▶ **Waterwheel**
Try designing your own simple waterwheel. This model has plastic blades fixed onto a cork, with a wooden dowel as an axle.

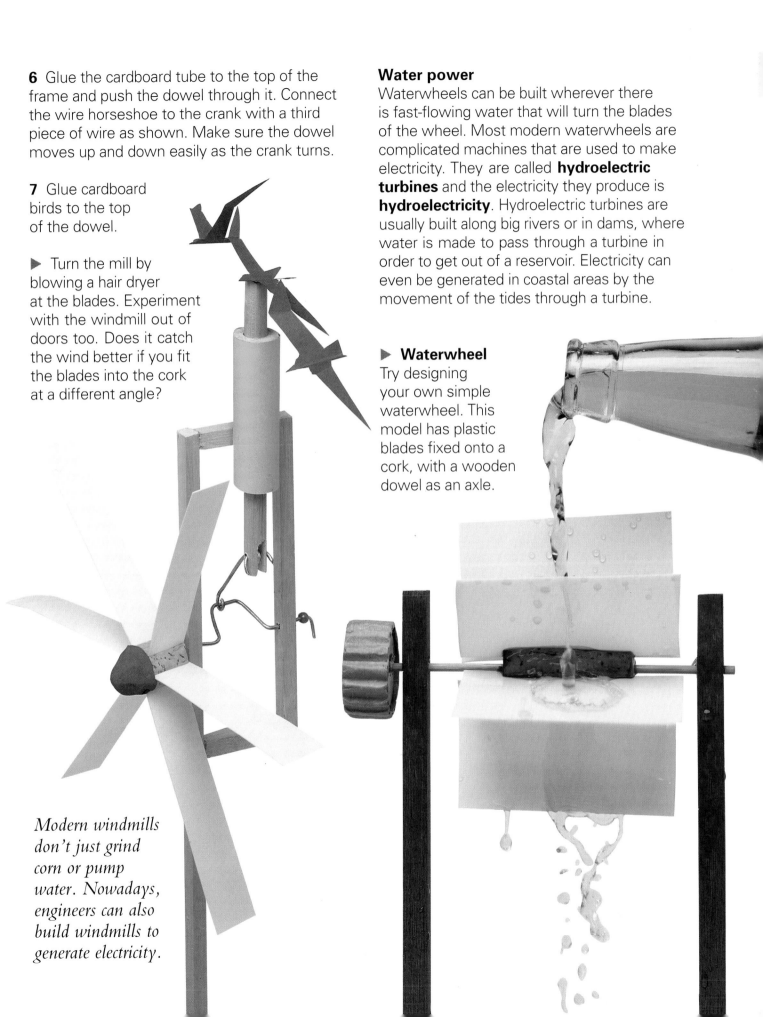

Rubber is an amazing material. You can stretch a rubber band to twice or three times its original length and it immediately springs back into shape when you let go. The stretched rubber stores energy. You can use this elastic energy to flick the band across the room, to make a catapult, or even to power model cars, boats, and planes.

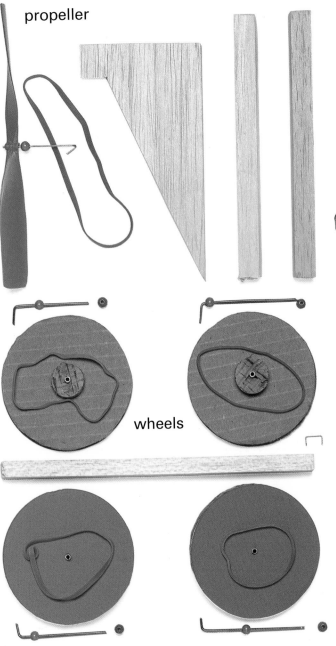

propeller

wheels

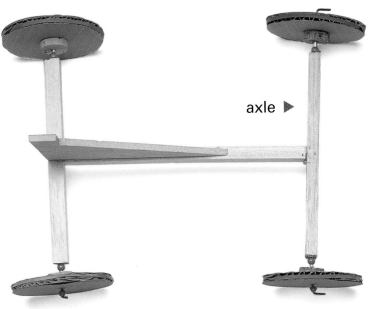

axle ▶

## You will need

balsa wood
a thick needle                     beads
heavy cardboard                 a cork
wire or thin nails                  a stapler
a plastic propeller               a paper clip
thick rubber bands             a craft knife
very short pieces of thin metal tubing

## MAKE it WORK!

This roadster uses energy stored in a twisted rubber band to turn a propeller. The propeller pushes against the air, driving the car forward.

The roadster's cardboard wheels are difficult to make, so you might want to buy ready-made plastic wheels from a model shop.

**1** Cut out the balsa wood parts of the roadster's frame and glue them together as shown.

**2** When the glue is dry, use a thick needle to make a hole in the frame for the propeller shaft.

**3** Unbend a paper clip and slip the model propeller onto it. Add a small bead on each side of the propeller.

**4** Push the paper clip through the hole in the frame, and bend both ends back. Hook one end of the rubber band over the paper clip and staple the other end to the front of the roadster's frame.

**5** Make the wheels. Cut out four disks of corrugated cardboard and four slices of cork. Glue a slice of cork to the center of each cardboard disk.

**6** Poke a small hole through the middle of each wheel and reinforce it with a very short length of metal tube as shown.

**7** Loop a rubber band around the rim of each wheel to make a tire.

**8** Fix a wheel to the end of each axle with a piece of bent wire or a thin nail. Put a bead on either side of each wheel so it can spin freely.

**9** Wind up the propeller and let your roadster go.

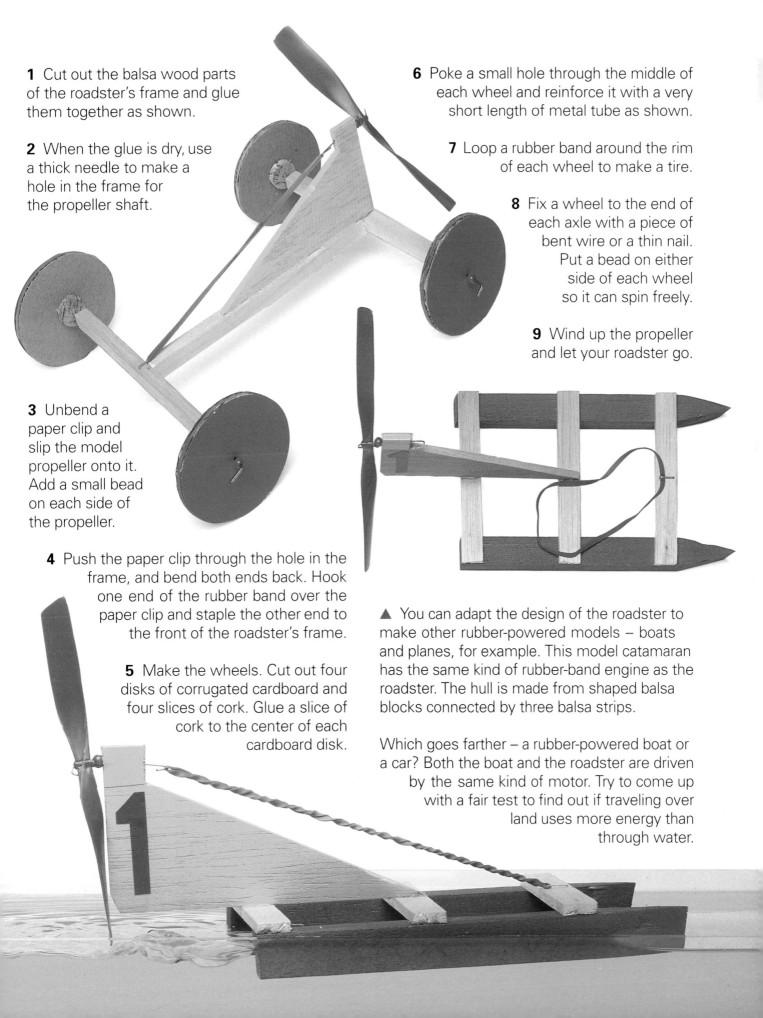

▲ You can adapt the design of the roadster to make other rubber-powered models – boats and planes, for example. This model catamaran has the same kind of rubber-band engine as the roadster. The hull is made from shaped balsa blocks connected by three balsa strips.

Which goes farther – a rubber-powered boat or a car? Both the boat and the roadster are driven by the same kind of motor. Try to come up with a fair test to find out if traveling over land uses more energy than through water.

For centuries, engineers dreamed of making machines that could fly like birds, but they didn't succeed until less than a hundred years ago. Because the force of gravity is so strong and air is so thin, a plane cannot get up off the ground unless it has a powerful engine and is very light for its size. A steam-powered plane carrying sacks of coal would never fly.

**You will need**

| | |
|---|---|
| a drill or awl | strong glue |
| a plastic propeller | a craft knife |
| thin cardboard and balsa wood | a rubber band |
| two paper clips and a bead | wire and a cork |

**1** Cut the wing, tail plane, and rudder from cardboard.

**2** Make the fuselage. Cut two strips of balsa wood 10 inches long and two strips 2 inches long. Glue them together as shown.

**3** Use a thin drill or awl to make a small hole through each end of the fuselage.

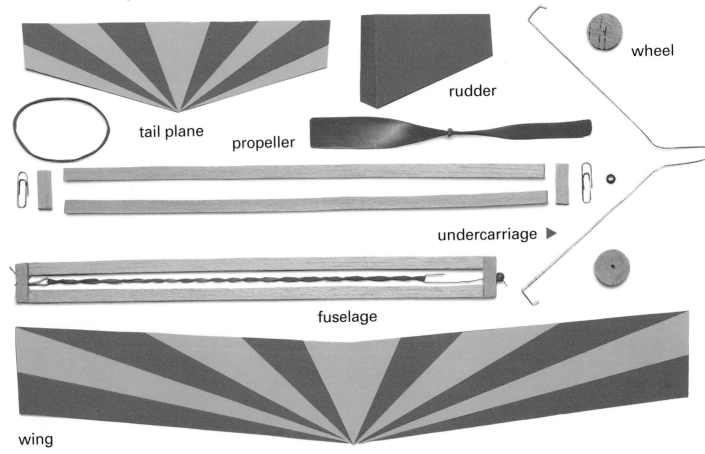

wheel

rudder

tail plane

propeller

undercarriage ▶

fuselage

wing

## MAKE it WORK!

Rubber is light, so it makes a good engine for a model plane. This model stays in the air for just a few seconds, but you'll find it travels farther than the roadster or the boat. Pushing through air is much easier than moving through water or rolling across the ground – so air travel can be more **efficient** than going by land or sea.

**4** Make two hooks out of paper clips. Loop the rubber band over them and fit one hook in each hole in the fuselage as shown.

**5** At the tail end, bend the hook all the way around and tape it firmly so that it won't move. At the propeller end, add a bead, so that the hook can spin freely.

**6** Attach the propeller and bend the paper clip hook around to hold it in place.

**7** Glue the wings, tail plane, and rudder in place.

**8** Bend the wire to make the undercarriage shape and push on slices of cork to make wheels. Glue the wire onto the fuselage.

**9** The plane is now complete – but before its first flight you must make sure that it balances well. Rest the wing tips on your fingers. If it does not balance, add small pieces of modeling clay to the nose or the tail until it is level.

### Test flight

Wind up the propeller until the band is twisted tightly. Hold the plane just behind the wing and launch it gently into the air.

If your plane nose dives, add more weight to the tail. If the plane stalls (the nose tips up and the plane slows down) add more weight to the nose.

*Compared to gasoline, rubber doesn't store much energy, so a rubber-band motor only powers short flights. But in 1979 one model flew for over 52 minutes – a world record!*

When water boils it changes into steam. The steam needs more space than the water, so it pushes against the things around it. The inventor of the first steam engine probably saw steam pushing the lid up off a boiling kettle and realized that this power could be used to push pistons and turn wheels.

## MAKE it WORK!

These model steam boats are powered by a candle. The candle's flame boils water inside a thin metal tube. Puffs of steam squirt from one end of the tube and push the boat along. More cold water is then sucked back into the tube, to replace the steam that has been puffed out.

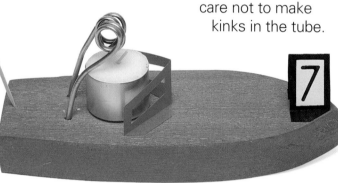

## ▲ Bending the metal tube

To make a steam boat, you will need to twist a piece of brass or copper tubing. The easiest way to bend the tube is by winding it around a length of dowel. Make the bends slowly, taking care not to make kinks in the tube.

## You will need

a toothpick
a small candle
cardboard and glue
a piece of soft brass or copper tubing
a length of flexible plastic pipe that will fit neatly over the metal tubing

balsa wood
a wooden dowel

**1** Cut the balsa wood into the boat shape.

**2** Twist the metal tube as shown above.

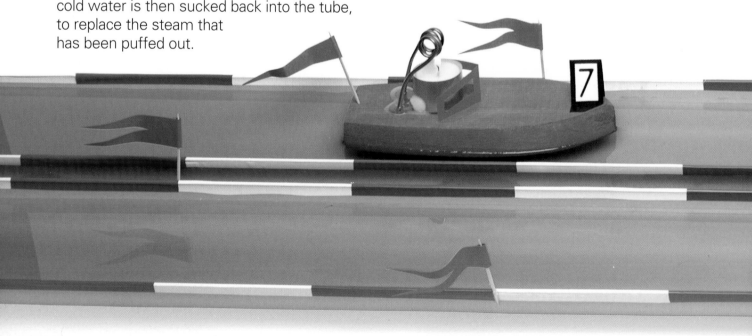

**3** Ask an adult to help you push the two ends of metal tube through the balsa wood.

**4** Cut the flag, windshield, and number plate out of cardboard, and fold and glue them in place. Then glue the candle on the boat as shown.

**5** Push the plastic pipe over one end of the metal tube, and float the boat on the water.

**6** Suck some water into the metal tube through the plastic pipe. When the tube is full, pull off the plastic tube, taking care not to lift the boat out of the water. Light the candle and watch your boat go!

*The first steam engines did work that used to be done by horses, so an engine's strength was measured in horsepower. A 10 horsepower engine could do the work of ten horses. Today, even a small car engine has more than 50 horsepower. But our steam boat has only the power of a small insect!*

### ▼ Steam-boat race

You could hold a race between two steam boats along lengths of plastic gutter that are filled with water.

What would you call a machine that repeats the same movement, hour after hour? It's a clock! A clock moves in a very regular way, counting out the passing minutes as it goes.

## MAKE it WORK!

Building a clock isn't easy – most machines tend to slow down or to get faster as they work. A machine that works steadily needs careful engineering to control the speed. Try it for yourself with this marble-operated clock!

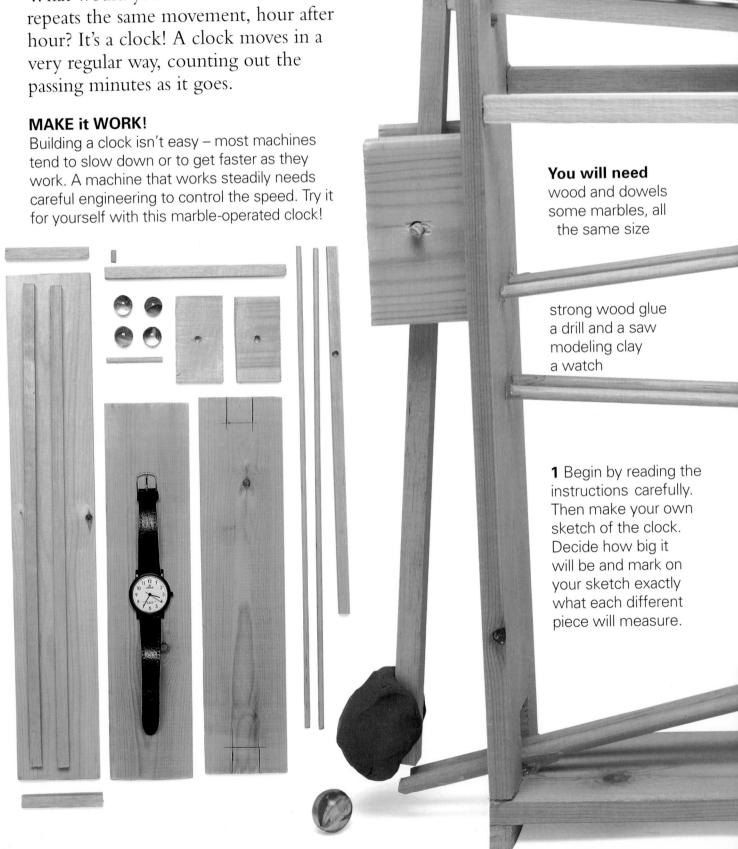

**You will need**
wood and dowels
some marbles, all
  the same size

strong wood glue
a drill and a saw
modeling clay
a watch

**1** Begin by reading the instructions carefully. Then make your own sketch of the clock. Decide how big it will be and mark on your sketch exactly what each different piece will measure.

**2** Ask an adult to help you cut the pieces of wood needed to make the clock's frame. Cut a slot at the top and bottom of one side piece in the positions shown.

**3** Firmly glue together the base, the side pieces, and the two top crossbars.

**4** Cut eight lengths of dowel to make slopes for the marbles to run down. Measure the dowels carefully, so that they are all slightly longer than the crossbars.

**5** Glue the slopes in place. Adjust them carefully, so that the gap between the two dowels is smaller at the top end and gets wider farther down. At the bottom of each slope, the gap should be just wide enough for the marble to drop through and onto the slope below.

**6** Cut two pieces of wood to make the lever arm. Drill a hole a third of the way down the longer piece. Glue both pieces together as shown, adding a small wooden stop to the end of the shorter piece.

**7** Drill a hole in each of the lever supports and glue them to the frame. Fit the lever using a short dowel peg. Make sure the arm can swing freely.

**8** Start a marble on the run. At the bottom, it should knock the lever just hard enough to release the next marble.

Use a watch with a second hand to time how long each marble takes to run from top to bottom.

**Adjusting the lever**
You'll probably need to adjust the lever to get it to work well. If more than one marble is released at once, add a modeling clay weight to the bottom of the lever.

A modern electronic clock is very accurate and reliable. It will run for over a year on just one tiny battery and will lose or gain no more than a few seconds in that time. Early clocks were much cruder.

All the clocks on this page have been used in the past. How accurate do you think they are? Try making them and test them against your own watch.

## Sand clocks

We still use sand clocks – sand running through an egg timer measures the minutes needed to cook a perfect boiled egg.

Make a cardboard funnel and fit it into the neck of a bottle. Fill the funnel with dry sand and see how long it takes to run through.

▶ Using a watch, make a scale on the sand clock. Mark the level of the sand at regular intervals (for example, every ten seconds). Are all of the marks evenly spaced? If not, why do you think the spacing changes?

## Candle clocks

Monks in the Middle Ages often used candles to measure the time. With the help of an adult, test a candle yourself to see how far it burns in an hour. Make hour marks along the rest of the candle with tape. Then mark the half and quarter hours too. How accurate is this clock?

## Water clocks

Do you ever lie awake at night listening to a dripping faucet? The drips are sometimes so regular you can guess exactly when the next one is coming. The ancient Chinese used dripping water to invent elaborate water clocks.

### To make a water clock you will need

tape
a glass
a straw
modeling clay
a large wooden bead
an old plastic container

**1** Make a scale by marking the straw with tape. Attach the straw to the base of the glass with modeling clay.

**2** Slip the bead over the straw.

**3** Make a small hole in the bottom of the container. Then fill up the container with water and hold it over the glass.

**4** As water drips into the glass, the bead rises up the scale. If the water runs too slowly, make a larger hole. If it is too quick, tape over the hole to make it smaller.

*More than three hundred years ago, the great Italian scientist, Galileo Galilei, sat in the cathedral at Pisa watching a hanging lantern swing back and forth. He realized that the lantern was a **pendulum**, and that each of its swings took exactly the same amount of time.*

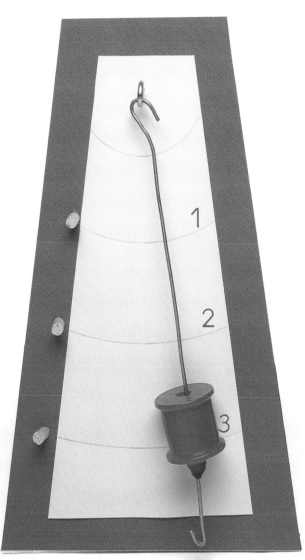

### ▲ Pendulum clock

A swinging pendulum is a good way of marking time. You can make a pendulum with a thread spool, modeling clay, and wire. Hang the spool from a hook and use your watch to investigate how the time taken by each swing changes as you move the spool up and down the wire.

Can you design any other machines of your own for measuring time?

**Axle**  A rod or shaft around which wheels and gears turn.

**Ball bearings**  Ball bearings help to reduce friction when a wheel turns around an axle. The steel balls are placed so that they roll between the wheel and the axle.

**Cam**  An oval shaped wheel, or a wheel whose axle does not go through the center. It is used to change a turning movement into an up-and-down movement.

**Compressed**  If something is compressed, it is squashed together by force. Compressed air stores energy that can be used to propel a model rocket.

**Drive belt**  A loop of rubber or other material that carries power from one pulley to another.

**Drive chain**  A loop of chain that does the same job as a drive belt. The links of the chain fit around the teeth on gears called sprockets. A bicycle chain is a drive chain.

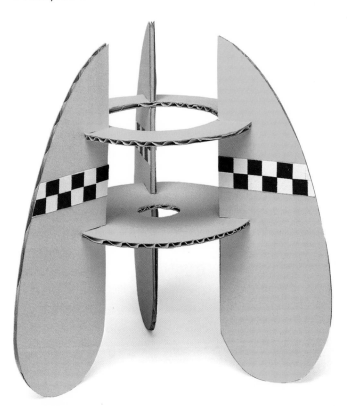

**Efficient**  An efficient machine does its job without wasting energy. Oil, for instance, helps many machines to run smoothly and efficiently by reducing the amount of friction between moving parts.

**Effort**  The force (a push or a pull) needed to work a lever or a pulley.

**Energy**  When something has energy, it has the ability to make other things move and change. People use the energy stored in their muscles to push and pull loads. Engines use the energy stored in fuels such as gasoline.

**Engine**  A machine that uses the energy from a fuel, such as coal or gasoline, to do work like lifting loads or turning wheels.

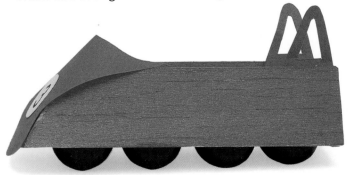

**Engineers**  Engineers use scientific knowledge to invent and make things. Besides machines, they design roads, bridges, and buildings.

**Force**  A push or a pull used to lift something, start it moving, or hold it in place against another force such as gravity.

**Friction**  A dragging force that keeps things from sliding across each other smoothly. Rough surfaces, such as sandpaper, cause more friction than smooth surfaces, such as ice.

**Fossil fuels**  Coal, oil, and natural gas are fossil fuels. They are formed from the remains of ancient plants and animals that have been buried for thousands of years beneath layers of mud and rock.

**Fulcrum** The hinge or pivot around which a lever turns.

**Gears** Toothed wheels that link together and carry turning movement from one place to another. Gears are also used to change the speed and direction of movement.

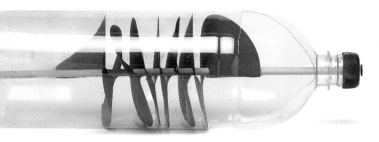

**Gravity** The force that causes objects to fall toward Earth, making them feel heavy.

**Gyrocompass** An instrument that always points in the same direction because it contains a spinning gyroscope.

**Hub** The center of a wheel. The axle usually goes through the hub.

**Hydroelectricity** Electricity produced by the energy of flowing water.

**Hydroelectric turbine** A machine turned by flowing water, which is connected to an electricity generator.

**Lever** A rod or bar that rests on a pivot. A load at one end of the lever can be lifted by applying an effort at the other end.

**Load** The weight or force that is moved by applying effort to a lever or pulley.

**Mesh** The teeth on two gears are meshed when they fit together. One gear turns and its teeth mesh with the teeth on the second gear and make it turn too.

**Pendulum** A hanging weight that swings back and forth because of the force of gravity.

**Pivot** A hinge or balance point around which something turns.

**Pneumatic** Pneumatic machines are driven by compressed air.

**Pollution** Waste or garbage that damages the natural world.

**Projectile** A missile thrown by a catapult or fired from a cannon.

**Pulley** A wheel turned by a rope or drive belt. It changes the direction of a force or carries it from one place to another.

**Pulley blocks** Two or more pulleys joined together to lift a large load with a small effort.

**Range** The distance traveled by a projectile between the firing and landing points.

**Rotate** To turn around, like a wheel on an axle.

**Sprockets** Toothed wheels that are usually connected by a drive chain.

**Trajectory** The curved path that a projectile follows as it travels through the air.

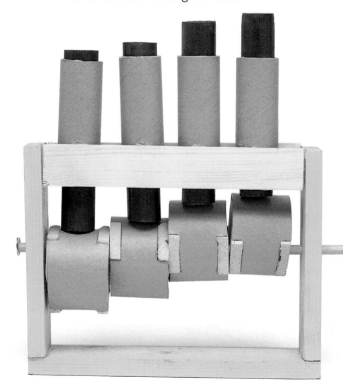